THE WEATHER FORECAST

ICE

A Crabtree Roots Book

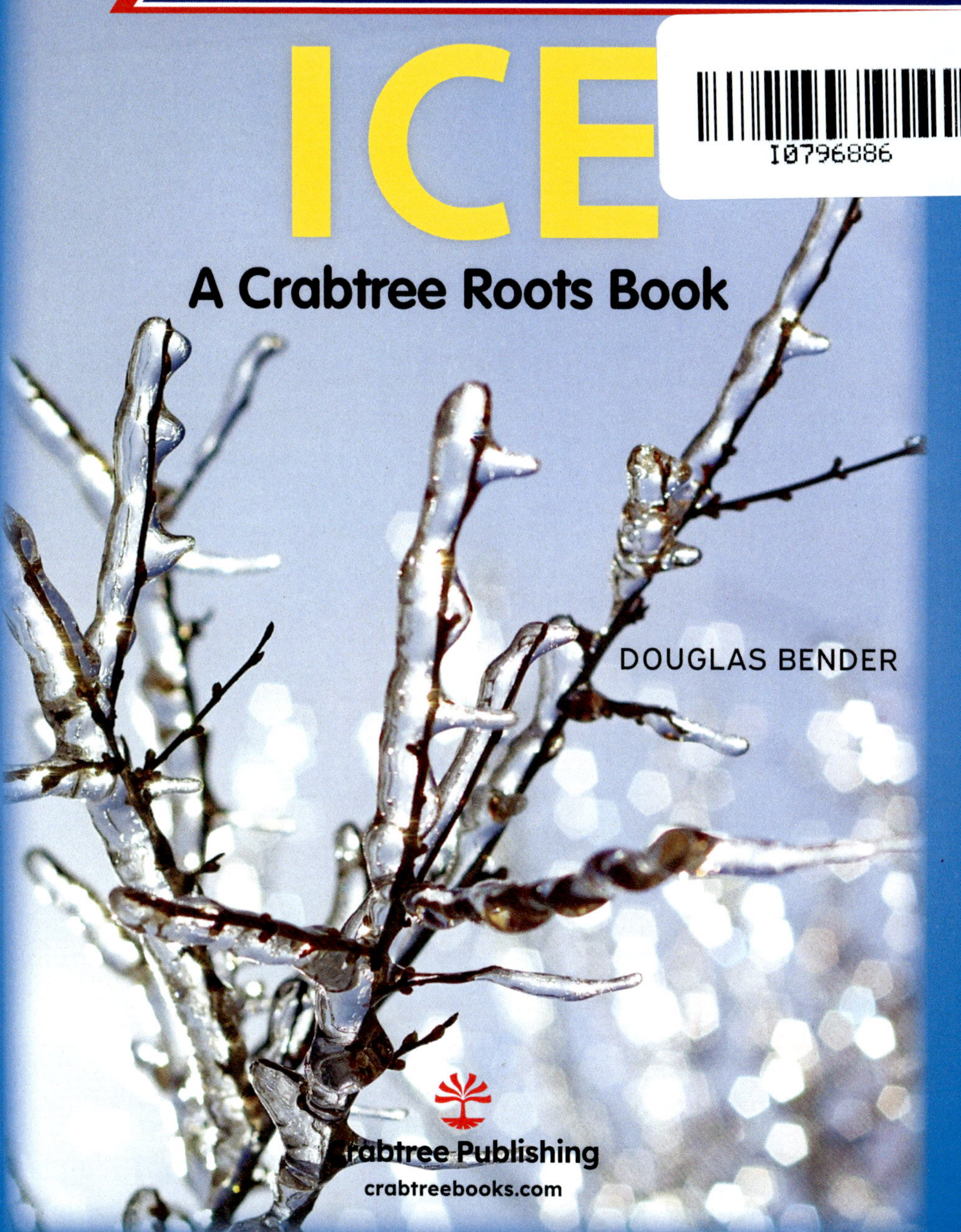

DOUGLAS BENDER

Crabtree Publishing
crabtreebooks.com

School-to-Home Support for Caregivers and Teachers

This book helps children grow by letting them practice reading. Here are a few guiding questions to help the reader with building his or her comprehension skills. Possible answers appear here in red.

Before Reading:

- What do I think this book is about?
    - *I think this book is about icy weather.*
    - *I think this book is about what icy weather looks like.*

- What do I want to learn about this topic?
    - *I want to learn where ice comes from.*
    - *I want to learn what ice is made of.*

During Reading:

- I wonder why...
    - *I wonder why pieces of hail can be big or small.*
    - *I wonder why water changes into ice.*

- What have I learned so far?
    - *I have learned that water can change into sleet.*
    - *I have learned that ice can form on land and in clouds.*

After Reading:

- What details did I learn about this topic?
    - *I have learned that it must be cold for water to change into ice.*
    - *I have learned that ice can form on land, in the air, and in clouds.*

- Read the book again and look for the vocabulary words.
    - *I see the word* ***hail*** *on page 6 and the word* ***frost*** *on page 13. The other vocabulary words are found on page 14.*

Brr! It is **icy** out!

Ice forms from water.

Ice can form in **clouds** too.

There is **hail** coming down.

Hail can be big
or small.

Water can also change into **sleet**.

Air and water can make **frost**!

Word List

Sight Words

and	into	out
big	is	small
can	it	there
from	make	too
in	or	

Words to Know

clouds

frost

hail

ice

icy

sleet

38 Words

Brr! It is **icy** out!

Ice forms from water.

Ice can form in **clouds** too.

There is **hail** coming down.

Hail can be big or small.

Water can also change into **sleet**.

Air and water can make **frost**!

ICE

Written by: Douglas Bender
Designed by: Rhea Wallace
Series Development: James Earley
Proofreader: Janine Deschenes
Educational Consultant: Marie Lemke M.Ed.

Photographs:
Shutterstock: Jaromir Chalabala: cover; Plus69: p. 1; Yekatseryna Netuk: p. 3, 14; Jake Hukee: p. 4; Ondra Vacek: p. 5, 14; Ikebana Art-studio: p.6-7, 14; hayim.bevzv: p. 9a; roseberry3; 9b; Evannovostro: p. 11, 14; Kimson: p. 12, 14

Crabtree Publishing

crabtreebooks.com 800-387-7650

In Canada: We acknowledge the financial support of the Government of Canada through the Canada Book Fund for our publishing activities.

Printed in Canada/092023/CPC20230905

Published in Canada
Crabtree Publishing
616 Welland Avenue
St. Catharines, Ontario
L2M 5V6

Published in the United States
Crabtree Publishing
347 Fifth Avenue
Suite 1402-145
New York, NY 10016

Hardcover 978-1-4271-5933-5
Paperback 978-1-4271-5939-7
Ebook (pdf) 978-1-4271-3377-9
Epub 978-1-4271-3437-0
Read-along 978-1-4271-5957-1
Audio book 978-1-4271-5963-2

Library and Archives Canada Cataloguing in Publication

Title: Ice / Douglas Bender.
Names: Bender, Douglas, 1992- author.
Description: Series statement: The weather forecast | "A Crabtree roots book".
Identifiers: Canadiana (print) 20210181435 | Canadiana (ebook) 20210181443 | ISBN 9781427159335 (hardcover) | ISBN 9781427159397 (softcover) | ISBN 9781427133779 (HTML) | ISBN 9781427134370 (EPUB) | ISBN 9781427159571 (read-along ebook)
Subjects: LCSH: Ice—Juvenile literature. | LCSH: Weather—Juvenile literature.
Classification: LCC QC926.37 .B46 2022 | DDC j551.31—dc23

Library of Congress Cataloging-in-Publication Data

Names: Bender, Douglas, 1992- author.
Title: Ice / Douglas Bender.
Description: New York, NY : Crabtree Publishing, [2022] | Series: The weather forecast - a Crabtree roots book | Includes index.
Identifiers: LCCN 2021014547 (print) | LCCN 2021014548 (ebook) | ISBN 9781427159335 (hardcover) | ISBN 9781427159397 (paperback) | ISBN 9781427133779 (ebook) | ISBN 9781427134370 (epub) | ISBN 9781427159571
Subjects: LCSH: Ice--Juvenile literature. | Weather--Juvenile literature.
Classification: LCC QC926.37 .B46 2022 (print) | LCC QC926.37 (ebook) | DDC 551.57/8--dc23
LC record available at https://lccn.loc.gov/2021014547
LC ebook record available at https://lccn.loc.gov/2021014548